SACRED GEOMETRY

Originally published in Wales by Wooden Books Ltd. in 1998; first published in the United States of America in 2001 by Walker Publishing Company, Inc.

Published simultaneously in Canada by Fitzhenry and Whiteside, Markham, Ontario L3R 4T8

Printed on recycled paper.

Library of Congress Cataloging-in-Publication Data
Lundy, Miranda.
Sacred geometry / written & illustrated by Miranda Lundy.
p.   cm.
ISBN 0-8027-1382-3 (alk. paper)
1. Geometry, Plane.   I. Title.
QA445 .L86   2001
516'.05—dc21   2001026013

Printed in the United States of America

6   8   10   9   7

SACRED
GEOMETRY

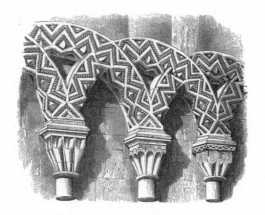

written and illustrated by
Miranda Lundy

Walker & Company
New York

This book is dedicated to the designers of the future.

My sincere thanks to my teachers Professor Keith Critchlow, John Michell, Dr. Khaled Azzam, Paul Marchant, Robin Heath, Michael Glickman, Dr. Stephan René, and Tony Ashton.

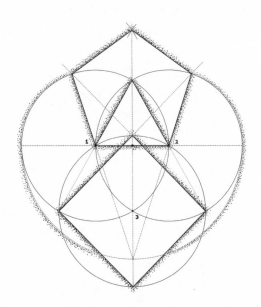

CONTENTS

Seats in Norwood's chantry, in Milton church, Kent.

INTRODUCTION

Sacred Geometry charts the unfolding of number in space. The basic journey is from the single point, into the line, out to the plane, through to the third dimension and beyond, eventually returning to the point again, watching what happens on the way.

This small book covers the elements of two-dimensional geometry—the unfolding of number on a flat surface. Another book in this series unfolds the three-dimensional geometrical story. This material has been used for a very long time indeed as one introduction to metaphysics. Like the elements of its sister subject, music, it is an aspect of revelation, a bright indisputable shadow of Reality and a creation myth in itself.

Number, Music, Geometry, and Cosmology are the four great Liberal Arts of the ancient world. These are simple universal languages, as relevant today as they have always been, and still found in all known sciences and cultures without disagreement. Indeed, one would expect any reasonably intelligent three-dimensional being anywhere in the universe to know about them in much the same way as they are presented here.

I do hope you enjoy this little volume and I urge further reading in this series for a deeper picture. With many thanks to the editors at Wooden Books.

Penzance, June A.D.2000

POINT, LINE, AND PLANE
none, one, and two dimensions

Begin with a sheet of paper. The point is the first thing that can be drawn on it. It is without dimension and is not in space. Without an inside or an outside, the point is the source for all that now follows. The point is represented as a small circular dot.

The first dimension, the line, comes into being as the one emerges into two principles, active and passive (*below*). The point chooses somewhere "outside" itself—a direction. Separation has occurred and the line comes into being. A line has no thickness, and it is sometimes said that a line has no end.

Three paths now become apparent (*opposite*):

1. With one end of the line stationary, or passive, the other is free to rotate and describe a circle, representing Heaven.

2. The active point can move to a third position equidistant from the other two, thus describing an equilateral triangle.

3. The line can produce another which moves away until distances are equal to form a square, representing Earth.

Three forms—circle, triangle, and square—have manifested. All are rich in meaning. Our journey has begun.

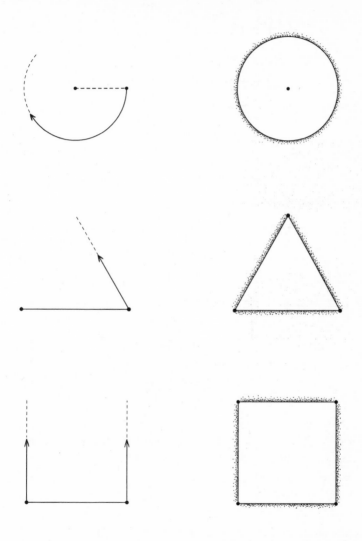

SPHERE, TETRAHEDRON, AND CUBE
from two to three dimensions

Although this book concerns itself primarily with the plane, the three "ways" are here taken one step further:

1. The circle spins to become a sphere. Something circular remains essentially circular (*opposite, top*).

2. The triangle produces a fourth point at an equal distance from the other three to produce a tetrahedron. One equilateral triangle has made three more (*opposite, center*).

3. The square lifts a second square away from itself until another four squares are formed and a cube is created (*opposite, bottom*).

Notice how the essential division into circularity, triangularity, and squareness from the previous page is preserved.

The sphere possesses the smallest surface area for its volume of any possible three-dimensional solid, whereas, among regular solids, the tetrahedron has the most.

The tetrahedron and the cube are two of the five Platonic solids (*see page 12*) and represent the ancient elements of fire and earth. The other three perfect solids are the octahedron (made of eight equilateral triangles), the icosahedron (made of twenty equilateral triangles), and the dodecahedron (made of twelve pentagons).

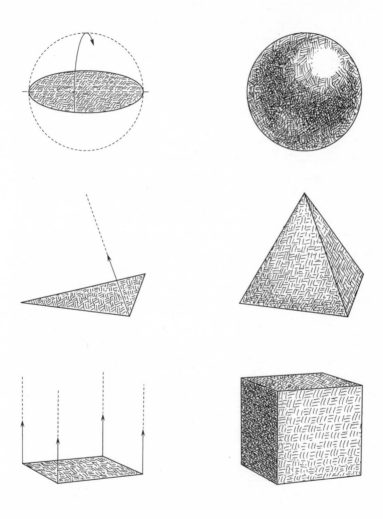

ONE, TWO, AND THREE
playing with circles

Get a ruler, a compass, something to draw with, and something to draw on. Draw a horizontal line across the page. Open the compass and place the point on the line. Draw a circle (*opposite, top*).

Where the circle has cut the line, place the compass point and draw another circle, the same size as the first. When one circle is drawn over another like this so that they intersect through each others' centers, an almond shape, the *vesica piscis,* literally "fish's bladder," is formed. It is one of the first things that circles can do. Christ is often depicted inside a vesica. Two equilateral triangles are defined within the vesica (*opposite, center*).

If a third circle is added on the other side of the first circle, then all six points of a perfect hexagon are defined.

Circles thus effortlessly produce perfect triangles and hexagons.

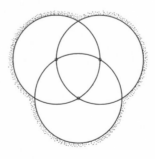

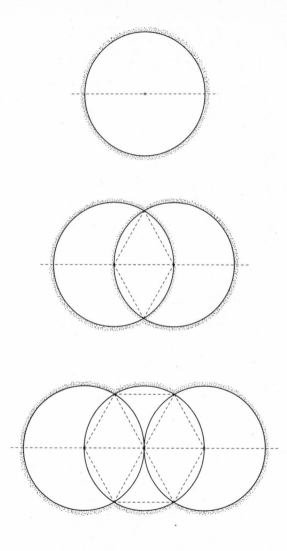

SIX AROUND ONE
or twelve or even eighteen

The six points of the hexagon give rise to the pattern shown below. Alternatively it can be drawn by "walking" a circle around itself—something most children have done at school, whether under instruction or just playing with compasses.

Looking at the diagram below, how might we locate the centers of the six outer circles? One way would be to lightly draw the six outer circles shown in the top diagram, opposite. Another way is to draw straight lines as shown in the lower diagram, opposite. Both ways work.

We can now see that six circles fit around one. We can push oranges, wineglasses, or tennis balls together to see it, yet it is extraordinary really. "Six around one" is a theme that the Old Testament of the Bible opens on, with the six days of work and the seventh day of rest. There is indeed something very sixy about circles.

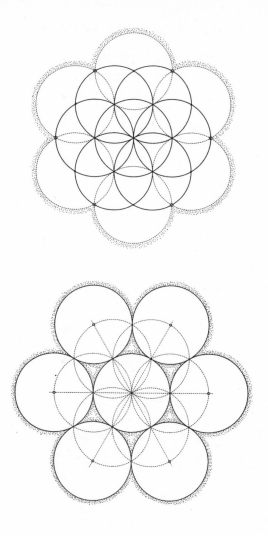

TWELVE AROUND ONE
how to draw a dodecagon

As one produces six, so six produces twelve. The arms of the six-pointed star extend to intersect the outer rims of the six circles to form a perfect overall division of space into twelve parts (*shown opposite*). The twelve-sided polygon is called a dodecagon, which means "twelve sided".

The dodecagon is also made from six squares and six equilateral triangles fitted around a hexagon. Can you see them all opposite?

Shown below is the three-dimensional version of the same story. A ball naturally fits twelve others around it so that they all touch the center and four neighbors. The shape made is called the cuboctahedron and is closely related to the tetrahedron and the cube we saw on page 5. Many crystals grow along these lines.

Twelve is the number that fits around one in three dimensions in the same way that six fits around one in two dimensions. The New Testament is a story of a teacher with twelve disciples.

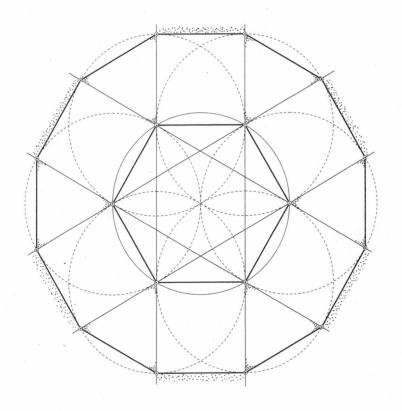

11

THE FIVE ELEMENTS
a brief foray into the third dimension

Although this book is primarily concerned with the first two spatial dimensions, it is important to our understanding of the number five to briefly explore the three-dimensional world.

There are just five regular three-dimensional solids. Each has equal edges, with every face of the solid the same perfect polygon. Every point is the same distance from the center. There are no others possible. They are known as the five Platonic solids, although they were recognized in the British Isles two thousand years before Plato; full sets of them have been found at neolithic stone circles in Aberdeenshire, Scotland.

The first solid is the tetrahedron, with four vertices and four faces (all equilateral triangles), traditionally representing the element of Fire. The second solid is the octahedron, made from six points and eight equilateral triangles, representing Air. The Cube is the third solid, eight vertices and six square faces, representing Earth. The fourth is the icosahedron, with twelve points and twenty faces of equilateral triangles, representing the element of Water. The fifth, and last, solid is the dodecahedron, which has twenty vertices, representing the mysterious fifth element of aether.

Notice how beautiful the dodecahedron is, and how it is made of twelve pentagons, perfect five-sided shapes.

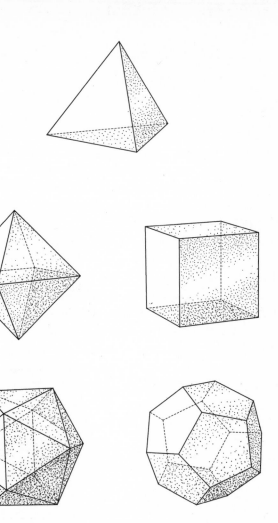

CIRCLING THE SQUARE
the marriage of heaven and earth

Traditionally, the circle is the shape assigned to the heavens, and the square to the Earth. When these two shapes are unified by being made equal in area or perimeter we speak of "squaring the circle," meaning that Heaven and Earth, or spirit and matter, are symbolically combined. Fivefold man exists between Heaven and Earth, and Leonardo da Vinci's image shows hand and foot positions that demonstrate both the circle and the square.

A circle and a square with exactly the same perimeter are shown below and, diagrammatically, opposite. Remarkably, if the Earth is fitted inside a square, then the equal perimeter circle defines the relative size of the Moon to 99.9 percent accuracy, as the space above the figure's head (*opposite*) and as the Moon drawn down to the Earth (*below center*). The Earth and the heavenly Moon thus square the circle. A standard compass construction for the square is also shown (*below left and right*).

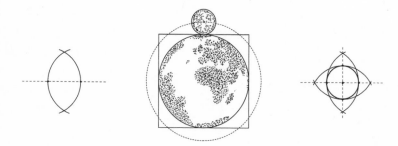

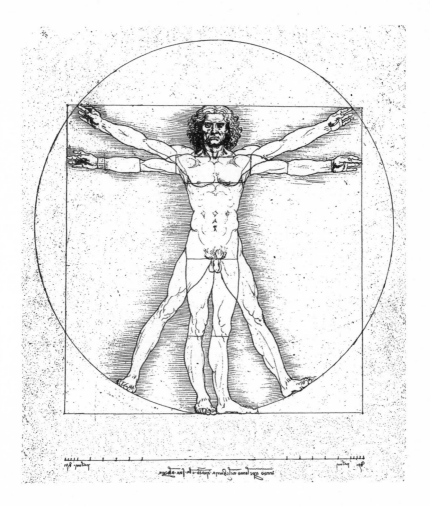

THE CANON
the numbers of the heavens and earth

If the Moon's radius is three, then the Earth's is eleven. The portal door of Gerum Church in Gotland, Sweden, (*opposite*) clearly shows a deliberate three-by-eleven proportion. Now, three times eleven is thirty-three, and Irish and Norse myths abound with tales of thirty-three warriors. Jesus dies at this age. It is also the case that, from any given place on Earth, the Sun takes thirty-three years before it rises exactly over the same point on a distant horizon. Seven works with both three and eleven as the Earth's tilt is seven in three, and a good value for π (*pi—the ratio of a circle's circumference to its diameter*) is 22/7.

Another important marriage is between five and eight. Shown below is a very close proportional agreement. In both diagrams the inner circle could be the size or orbit of Mercury if the outer circle is taken as being the size or orbit of Earth. Venus, although not shown, sits between Mercury and the Earth and draws a huge pentagram around us every eight years.

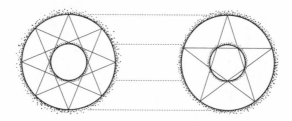

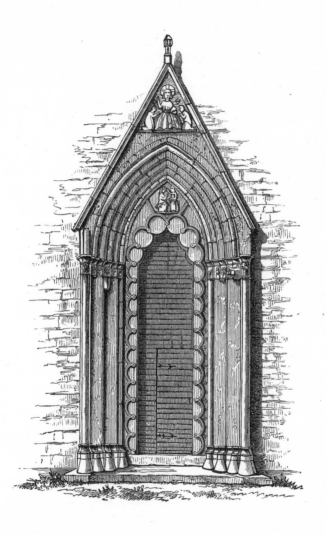

PYRAMID PIE
a marriage of everything

There is perhaps no more famous a geometric object on Earth than the Great Pyramid at Giza in Egypt with its strange passageways and enigmatic chambers. The five diagrams opposite show:

1. The square of the height as an area equal to that of each face.

2. The *golden section* in the pyramid, ϕ = 1.618 (*see page 24*).

3. *Pi* in the pyramid. Pi, or π, defines the ratio between a circle's circumference and its diameter (3.14159...).

4. The pyramid squaring the circle (*see page 14*).

5. A pentagram defining the "net" for the pyramid—cut it out!

Geometry means "Earth measure." The Pyramid functions as a ridiculously accurate sundial, star observatory, land surveying tool, and repository for weights and measures standards. Written into the design are highly accurate measurements of the Earth, detailed astronomical data, and these simple geometric lessons.

A 3-4-5 triangle fits the shape of the "King's chamber" (*below*) and also gives the angle of slope of the second pyramid at Giza.

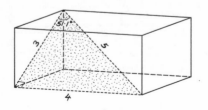

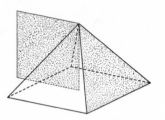

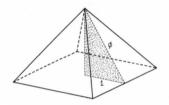

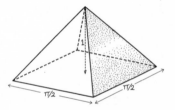

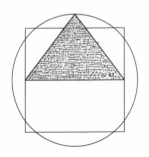

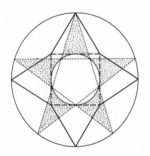

THE HALF AND THE THIRD
defined by triangles and squares

An equilateral triangle (*opposite, top left*), or two nested squares (*opposite, top right*) are similar in that the circle inside each of these figures is exactly half the size of the surrounding circle. This is a geometrical image of the musical octave, where a string length or frequency is halved or doubled.

Appropriately, the three-dimensional equivalent of the triangle, the tetrahedron, defines the next fractional proportion, one-third, as the ratio of the radius of the innermost sphere to that of the containing sphere (*opposite, bottom left*). Two nested cubes or two nested octahedra or an octahedron nested in a cube (*opposite, bottom right*) all produce one third too. The geometric third is musically equal to an octave plus a fifth in harmonic notation.

Thus *two* dimensions quickly define a *half*, and *three* dimensions a *third*. Another fascinating third is shown below.

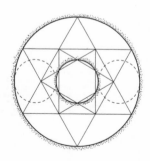

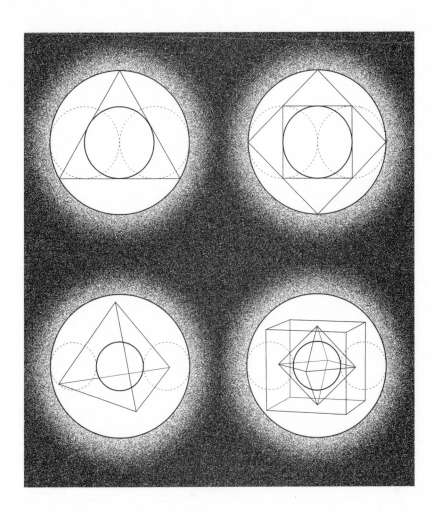

THE SHAPES OF SOUNDS
and three quarters

Geometry is "number in space," music is "number in time." The basic set of musical intervals is the elementary set of simple ratios, 1:1 (unison), 2:1 (the octave), 3:2 (the fifth), 4:3 (the fourth), and so on. The difference between the fourth and the fifth, which works out at 9:8, is the value of one whole tone. Musical *intervals*, like geometrical *proportions*, always involve two elements in a certain ratio: two string lengths, two periods (lengths of time), or two frequencies (beats per length of time).

We can see musical intervals as shapes by swinging a pen in a circle at one speed, and a table in an opposite circle at another speed, the device being called a Harmonograph. Shown opposite are two patterns from near-perfect intervals. The octave (*top*) draws as a triangular shape, the fifth (*bottom*) as a pentagonal form.

Two octaves, or a quarter, can be exactly defined by two triangles, four squares, or by a pentagon in a pentagram (*below*).

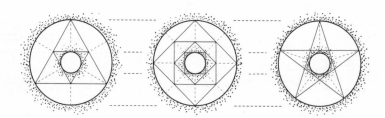

THE GOLDEN SECTION
and other important roots

A pentagram inside a pentagon is shown opposite. A simple knot, carefully tied in a ribbon or strip of paper and pulled tight and flattened out makes a perfect pentagon. Try it some time!

In the main diagram opposite you can see that pairs of lines are each broken differently. The length of each such pair of lines is in the *golden section* ratio, 1:ϕ, where ϕ, or *phi,* can be either 0.618 or 1.618 (more exactly .61803399. . .).

Importantly, ϕ divides a line so that the ratio of the lesser part to the greater part is the same as the ratio of the greater part to the whole. No other proportion behaves so elegantly around unity. For instance, $1 \div 1.618$ is 0.618 and $1.618 \times 1.618 = 2.618$. So one over ϕ is ϕ minus 1, and $\phi \times \phi$ is one plus ϕ!

The golden section is one of three simple proportions found in the early polygons (*opposite bottom*). With side lengths 1, a square produces an internal $\sqrt{2}$, a pentagram 1.618, and a hexagon $\sqrt{3}$. Although $\sqrt{2}$ and $\sqrt{3}$ are found widely in the animal, vegetable, and mineral kingdoms, ϕ appears predominantly in organic life and only rarely in the mineral world. All these proportions are employed in good design.

Neighboring terms in the Fibonacci series: 1, 1, 2, 3, 5, 8, 13, 21, 34, 55, 89, 144. . . (*add the preceding last two numbers to get the next*) increasingly approximate ϕ. For the keen $\phi = \frac{1}{2} (\sqrt{5}-1)$.

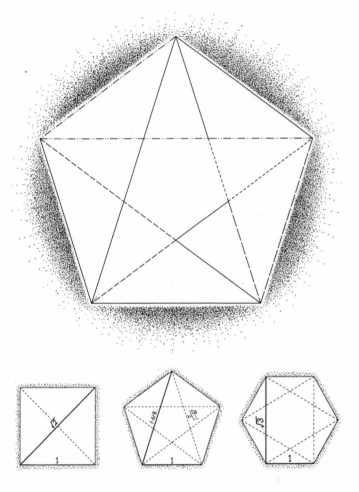

HOW TO DRAW A PENTAGON
and a golden section rectangle

The method of construction of a pentagram shown opposite is perfect and first appeared in the last book of Euclid's *Elements* two and a half thousand years ago.

Draw a horizontal line with a circle on it. Keeping the compass fixed, place the point at (1) and draw the vesica through the center of the circle. Now open the compass wide and draw arcs from (1) and (2) to cross above and below the circle. Use a straight edge to draw the vertical through the center of the circle. Next draw the vertical through the vesica to produce (3). With the point of the compass at (3) swing an arc down from (4) at the top of the circle to give (5). With the point at (4) swing through (5) to give two points of the pentagon. With the point of the compass on these new points in turn, swing from the top to find the last two points of the pentagon.

A golden section rectangle, widely used in architecture, is constructed from the midpoint of the side of a square (*below*).

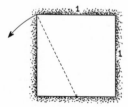
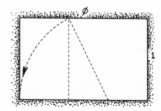

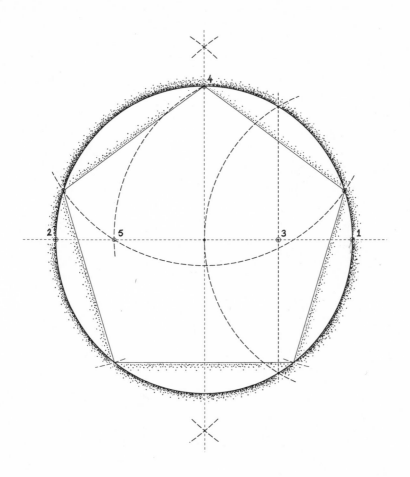

VARIOUS SPIRALS
and how to draw them

Spirals are marvelous forms that nature uses at every scale. Three have been selected for this book that give the impression of a spiral from multiple arcs of circles.

The first is the Greek Ionic volute shown top left. This is quite hard to draw, and the secret lies in the small key shown above it. The dotted lines in the main drawing show the radii of the arcs and give clues to the centers. It's not as confusing as it looks!

Regular spirals such as the one shown top right also need a key. This can simply be two dots, a triangle, a square, a pentagon, or a hexagon (*as shown*). The more points you have the more perfect the spiral will be. Simply draw the tiny key and open the compass wide to draw your first arc until it lines up with the next point of the key. Now move the compass point nearer to the arc, reduce the angle to fit, and draw the next section. It sounds much harder than it is—if you try it you will soon get the idea. The bigger the key the wider the coils. Now look at the Ionic volute key again. Can you see what is happening?

The bottom picture shows a golden section spiral, common throughout the natural world. A golden section rectangle (the Georgian front door and window) has the property that removing a square from it produces another golden section rectangle and this spiral uses that fact to create quarter arcs in each of the squares.

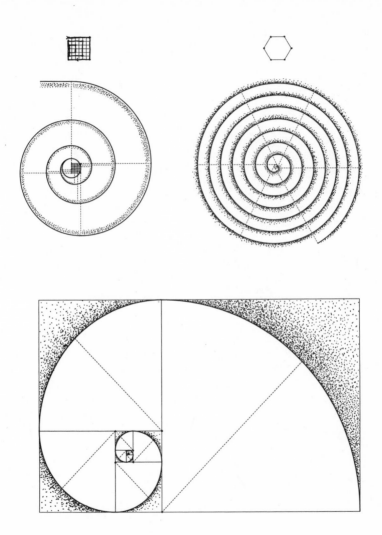

THE HEPTAGON
seven out of three

Divide a circle into six and draw the primary equilateral triangle. Find the midpoints (1) and (2) of the triangle's two upper arms and drop two lines down to give two points (3) and (4) on the base of the triangle, and two on the bottom of the circle. Finally, from the top, swing through the four points on the triangle to give the last four points of the seven on the circle.

Three and seven often work together (*see page 16*); a rectangle three across and seven high gives a diagonal that is the tilt of the Earth, or in many ancient pictures the tilt of a holy head.

Although it is nearly impossible to draw a precise heptagon using ruler and compasses alone, you can do it perfectly using seven equal rods or matchsticks (*shown below left*). This wedge is an *exact* fourteenth of a circle, so you need two of them for a one-seventh division. More ancient rough solutions use a cord with either six knots or in a loop with thirteen (*below center and right*).

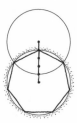

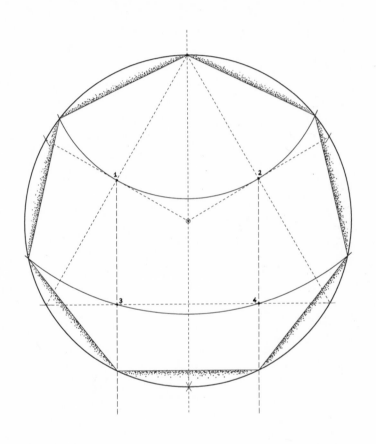

THE ENNEAGON
nines and magic squares

The construction shown opposite divides a circle into a near-perfect nine from an initial six-pointed star using three centers.

The digits of many special numbers add up to nine: 2,160 or 7,920 for instance, the diameters of the Moon and the Earth given in miles; 360 and 666 do, and so do all pentagonal angles like 36, 72, and 108. In fact, all multiples of nine add up to nine. Nine is three times three, or three "squared." Many tribal cultures speak of nine worlds, or nine dimensions.

Shown below is the basic eight-pointed star that can be drawn in a square. This simple device exactly divides a square into nine, sixteen, or twenty-five—the squares of three, four, and five.

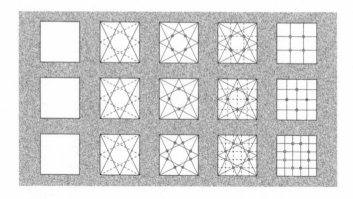

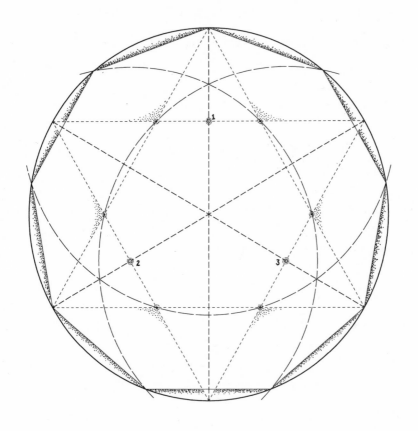

COIN CIRCLES
the structures that circles make

In both two and three dimensions, points or vertices can be regarded as circles or spheres. If you collect pennies, marbles, or peaches you can produce a variety of grids and learn a lot about how space arranges itself on the plane.

All the grids shown on the next four pages can also be drawn as arrangements of equal touching circles.

The most common pattern is the triangular repeat (*opposite, top left*). The bottom diagram opposite is another way of seeing overlapping dodecagons (*see page 11*).

Not for beginners is the extremely obscure fact that a sphere-point enneagon (*see previous page*) can contain two more spheres that exactly touch (*shown below*).

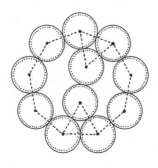

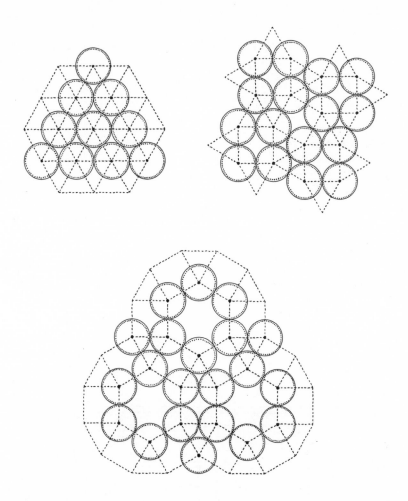

TILINGS

repeating patterns over an infinite surface

There are three regular tilings of the plane (*below*) and eight semi-regular (*opposite*). The top left and right grids opposite are left- and right-handed versions of the same pattern and count as one. All these can be drawn easily with a ruler and compass. Regular tilings are those where only one regular polygon is used to fill the plane, whereas semi-regular tilings allow for more than one type of polygon but insist that each meeting-place, point, or vertex is the same. For instance, in the center pattern opposite every vertex is a meeting of two hexagons and two triangles.

Some designs can be filled in further: As we saw on page 11, dodecagons are just made of hexagons, triangles, and squares, and hexagons can be made of triangles. Triangles and squares can go on to do the most amazing things together (*next page*). Octagons only tile with squares (*opposite, top center*). Pentagons do not fit together easily on the plane, preferring the third dimension (*see pages 12–13*). Heptagons and enneagons stand aloof.

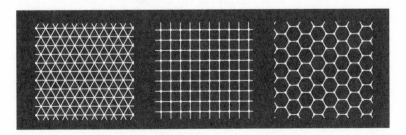

MORE TILINGS
the fourteen demi-regular grids

More than twenty demi-regular ways to tile the plane are shown on these two pages (where two vertex situations are permitted).

These tilings form the basis for pattern construction in many traditions of sacred and decorative art. They can be found underlying Celtic and Islamic patterns and, in the natural world, they appear as crystal and cellular structures. William Morris used them widely for his repeat wallpaper and fabric designs. Their uses are limited only by your imagination!

On page 39 we see one of these grids put to use.

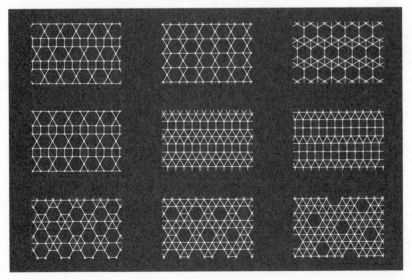

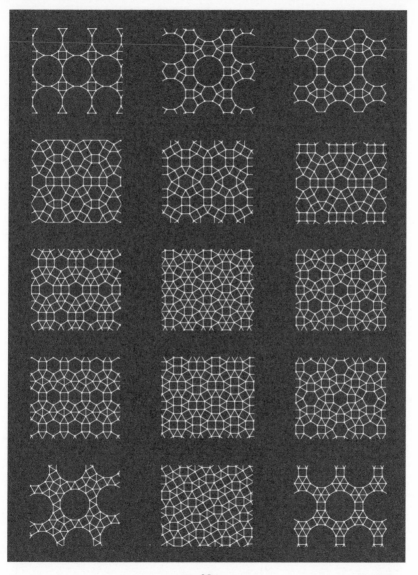

THE SMALLEST PART
reversible stencils and rotatable blocks

All the semi- and demi-regular grids can be reduced to a square or triangular unit that can then be reflected or rotated to recreate the whole pattern. Often these repeat triangles or squares are quite small. However, in practical applications, it is often easier to rotate a printing block or stencil than it is to reflect it—and in such cases one has to draw a larger stencil or carve a larger block.

The design shown opposite is based on one of the grids on the previous page. It is produced by rotation *and* reflection of the primary unit (*opposite, top and below right*).

Squares and equilateral triangles can both be halved to produce smaller triangular units (*below left*). One has to be careful doing this with the example shown opposite. Can you see why?

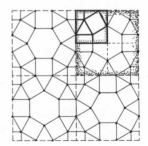

AN ISLAMIC DESIGN
stars are born from a subgrid

Islamic patterns speak of infinity and the omnipresent center.

For this pattern, start with six circles around one, developing a grid of overlapping dodecagons from triangles, squares, and hexagons (*see page 11, and page 37 bottom row, middle*) .

The key points now are halfway along the side of every polygon. These are joined in a special way and extended as shown below and opposite. Many beautiful patterns are sitting in every simple subgrid, just waiting to be pulled out.

The subgrids themselves are rarely shown in traditional art. They are considered part of the underlying structure of reality, with the cosmos, which means "adornment", overlaid.

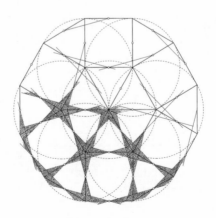

43

A CHURCH WINDOW
not far from the Isle of Man

A typical piece of church window masonry is shown opposite. It is a very beautiful design and the sort of thing that can be seen in many churches.

See if you can follow its construction from the diagram below, noticing how every detail is defined by the geometry. The outer circle is drawn first, divided into six, and the triangle drawn. The three touching circles follow, and notice in particular how they do not quite touch the outer circle that has produced them. A small circle that represents this then gives the width of the stone tracery itself.

The design speaks of the implicit trinity in unity.

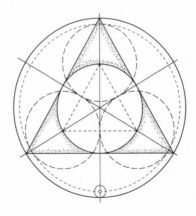

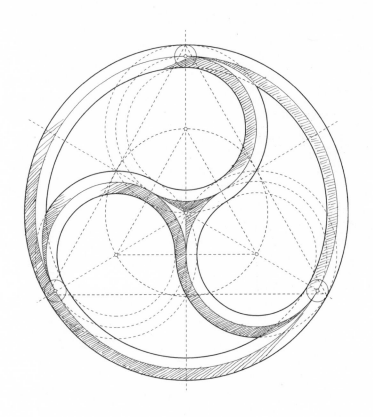

TREFOILS AND QUATREFOILS
and other church details

Everything is made of light, all matter is, and without matter there would be no sound. Atoms and planets arrange themselves in geometrical patterns. How profound then is a window, which allows the passage of light into an otherwise dark space.

The designs of church windows follow many rules, forms, and traditions, and some clues are given on these pages. The easiest to draw are the three quatrefoils (*bottom row below*).

The south window of Lincoln Cathedral is shown opposite, and below it three west windows, from the cathedrals at Chartres, Evreux, and Rheims. A balance is kept between line and curve.

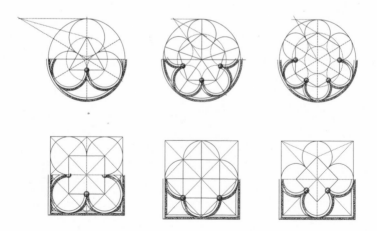

47

STONE CIRCLES AND CHURCHES
vesicas in action for over 4,000 years

Four stone circles are shown opposite with their consistent geometry as discovered by Professor Thom in the 1960s. On the left are Dinnever Hill and Cambret Moor, examples of the type-A flattened shape; on the right are Long Meg and Barbrook, two examples of type-B flattened stone circles. These two geometric shapes are found widely throughout the British Isles. The vesica-based constructions are shown below (*see page 6*).

Shown on this page is the plan of Winchester Cathedral. An interplay of simple triangular and square systems, *ad triangulum* and *ad quadratum*, underlies much western sacred architecture (*see top row page 21*). The vesica is central to church architecture.

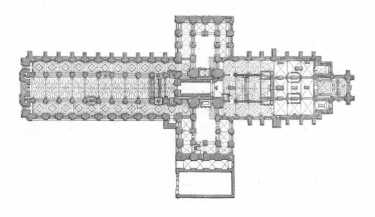

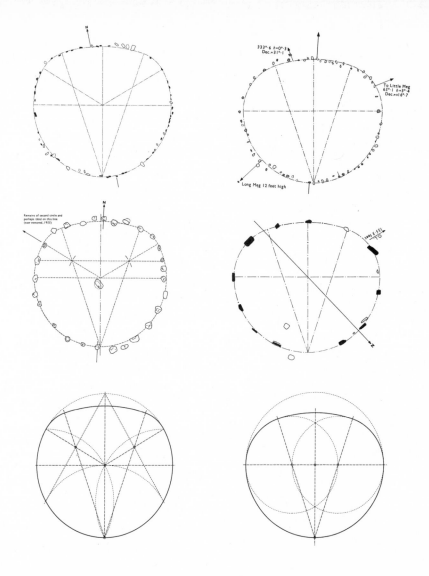

DELIGHTFUL ARCHES
how to draw a few of the many

Arches take remarkably similar forms all over the world, and a few are shown here. Living trees often make the best arches.

The top row opposite shows five two-centered arches. Their span has been divided into 2, 3, 4, 5, and 5 again. The straight dotted lines show the radii of their arcs. The heights of arches can vary, but for these five their heights are defined by a rectangle, which gives a musical interval, thus 2:3, 3:4, and so on (see *page 22*).

The second row of arches opposite are four-centered. The curve of the arch changes at a position given by the solid line. Ideas for defining their heights are also given.

The bottom two arches opposite are a horseshoe arch (which can also be pointed) and a pointed arch. The pointed arch seems to turn up ("the return") but the lines are actually dead straight.

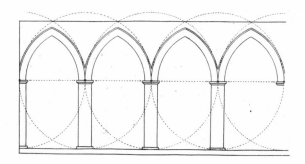

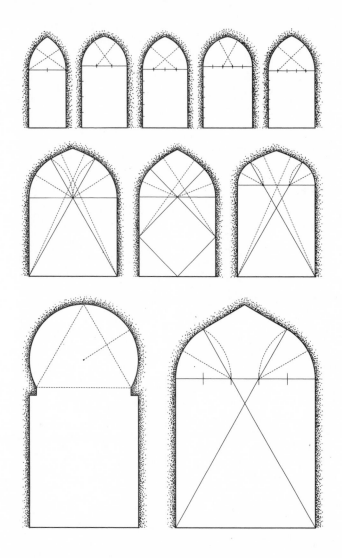

A CELTIC SPIRAL
Euclidean geometry in ancient Ireland

The design shown opposite comes from a four-inch bronze disk found on Loughan Island in Northern Ireland. It is an exceptionally beautiful example of the early Celtic style. As we have already seen with stone circles and arches, the seamless conjunction of multiple arcs can be highly aesthetic, and it perhaps reached its perfection in the early Celtic period.

Many early Celtic pieces show evidence of compasses, and drawing this disk required 42 separate compass-point positions! It is thought that the master artists who created these designs started with a basic geometric template, such as a touching circles pattern, then sketched their forms, and then returned to geometry to tighten everything up so that their curves were all *arcs*, sections of circles. In this way intuition and intellect worked together.

The bottom sequence of pictures shows how to plot arcs through points. The first diagram shows an arc centered on *c*. We want the arc to effortlessly change at *a* and then pass through *b*. What do we do? We draw the line between *a* and *b* and find its perpendicular bisector (*center diagram*). This cuts the *ac* line at a new point *o*, which then becomes the center we were looking for (*right diagram*). All of the curves in the Loughan Island disk are drawn in this way.

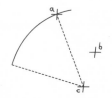

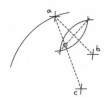

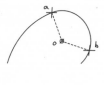

PENTAGONAL POSSIBILITIES
those phantastic phizzy phives

Although the pentagon does not tile on the plane, it does do various other things, which no good book on sacred geometry should omit to mention. One of these is shown opposite, where a "seed" pattern can be made to grow *from the center*. The pentagons leave spaces that are bits of pentagrams and vice versa. The design is riddled with examples of the golden section. Sample "seeds" are shown opposite below.

The mathematician Roger Penrose recently discovered the tiling shown below. Just two shapes tile to fill the plane. These patterns have been recently found to underly the nature of liquids. They are cross sections of higher-dimensional lattices.

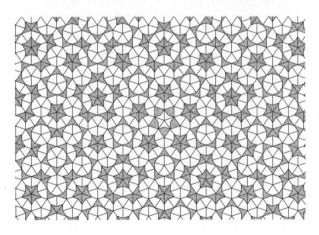

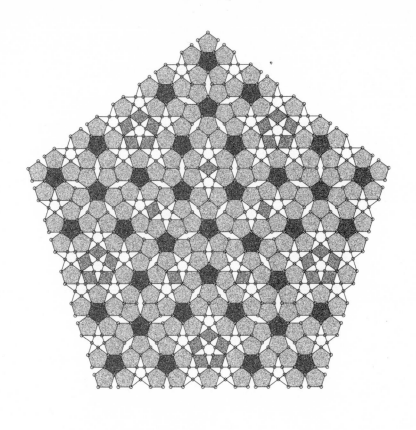

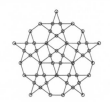

55

SEVENTEEN SYMMETRIES
from slide, spin, and mirror

The Arab alchemist Jabir ibn Hayyan, known in the West as Geber, regarded seventeen as the numerical basis of the physical world.

Using a very simple sample design, the next three pages explore the three basic operations of rotation, reflection, and sliding. These, combined with the three regular tilings, give seventeen patterns that are shown below, opposite, and on the next page.

This visual key can be very useful when creating repeats for fabric or pottery patterns (*see pages 40–41*). "Pattern," by the way, comes from the latin word *pater*, meaning "father," in the same way that *matrix* comes from *mater*, meaning "mother."

Remember, not all stencils can be turned over (reflected) without making a mess so choose your repeat units with care.

And staying on that rather practical note, this dense little book on one of the oldest subjects on Earth has now reached its end!